Lessons from Tesla:
Harnessing the Power of the Earth's Magnetic Field

Author: Michael C Ellis

Cover Design and Artwork: Michael C Ellis

©2019 Michael C Ellis

Disclaimer:
The author makes no claims, medical or otherwise. Radiant energy and wireless electricity has not been tested for efficacy, reliability, or safety by Underwriters Laboratory, the FCC, or any other government agency. If you attempt to use this technology you must obey all relevant laws and regulations in your territory (ie. radio interference, radiation standards, electrical safety and certification of operator, etc.). You operate such devices at your own risk.

Absolute Resistance Press

Monona, Wisconsin

Dedicated to every person fighting for freedom, peace, and enlightenment. I'm with you!

Also dedicated to my parents and my brother and sister in-law. See, I did something without screwing it up....

Preface

This book took half of a lifetime to write. I began researching energy and fringe topics when I was half my age, and I'm now 33 years old. It began when our parents got me my own computer for my room, so I could surf the internet undisturbed. I intuitively knew that our civilization required new energy technologies but little did I know that I would be writing a book about it. I began researching orgone technology first, mostly about Reich's work and following the modern stuff with Don Croft and the others. The orgone research community and those who write about their experiences with orgone technologies are essentially the ones who carried me through over the years, both artistically and intellectually. I also learned a lot from Borderlands, especially about Nikola Tesla and his electrical genius. I have to say that those two groups gave me the most information I needed to write this book. Tom Bearden and John Bedini both inspired me in electronics and physics and added the scholarly background I needed to relate this to my readers. Then I entered college in Electrical Engineering Technology to further my interests in energy physics and I finally developed the basic skills to start working with this stuff. I have to give credit to my professors at MATC for their support, even though they may have been somewhat close minded about the fringe stuff. They taught me the right way to go about it, and as you will see, I took that academic format to model one of my discussions on Tesla's radiant energy theory. I finally graduated with my BS in Art after going back to Edgewood for extra street cred, so thanks to my professors for giving me the skills to illustrate my books and write halfway decently. I even obtained the mystical powers only ordained to those anarchic folks that live in the shadows and print their own books. But alas, the internets make things much easier. At least I'll have a backup plan for the apocalypse.

Introduction

Nikola Tesla was a genius. Not only did he make cool lighting, burn up a power plant generator, and cause a small earthquake, he discovered how to harness the power of the planetary ecosystem. Not many people can agree on this, but not only did Tesla possess the technology to broadcast 'no cost' free energy, but 'no fuel-source' free energy. When we say 'free energy' we must conceded that all energy must have a source, ours is simply the dynamo that drives the planet, the sun, and the galaxy. We use magnetic energy.

Magnetic plasmas have been of interest lately since their discovery at MIT in a power transmission experiment. They have also been witnessed by NASA transferring energy between galaxies. It is well known that magnetic currents pool across the Earth's surface, not only at the poles, where the auroras are visible.

If you can move a compass, why can't you power an engine, or recharge your phone?

Part 1

The Experiment

Experiment to Demonstrate the Nature of Radiant
Energy Coupling, Negative Resistance,
and Transmission of Electricity with A Single Wire

Michael C Ellis, A.A.S., B.S.
2011

Updated on November 8th, 2019

1.1 Abstract

In this experiment, Nikola Tesla's observation that a novel form of electricity interacts with conductors during a DC impulse event was confirmed. A shorted LED, lengths of wire, and small inductance were used to demonstrate the properties of this event. Further experimentation demonstrated that this form of electrical energy is capable of affecting shorted and open circuits of various design, and does not use radio waves or capacitance to enter the shorted and open circuits. It was found that this energy is capable of creating current in any direction and is dependent upon the relaxation properties of the wires used when electricity is applied. This energy is more efficient at transferring power than any radio waves that were created by the spark event, which was also demonstrated.

2.1 Introduction

Nikola Tesla originated the idea of using a single conductor to transmit electricity to a remote load with both ends terminating on the conductor. This is a type of transmission line based circuit in which the signal or voltage is sent along the line without a return. Transmission is achieved by magnetic resonance via an open-ended transformer designed by Tesla, and tuned coils pick up that signal and power a load through a complementary transformer. Tesla observed that nodes of voltage form across any inductance attached to the conductor, and asymmetrical resonance (creating AC resonance with a DC, positive-offset wave) can be used for both power transfer and gain. This method does not rely on the ionization of air or the creation of light rays or radio waves, thus, ideally, large amounts of power can be transmitted without major losses. According to Tesla, in "The Transmission of Electrical Energy Without Wires As a Means of Furthering Peace":

"This invention, which I have described in technical publications, attempts to initiate, in a very crude way, the nervous system in the human body.... That electrical energy can be economically transmitted without wires to any terrestrial distance... it is practicable to distribute power from a central plant in unlimited amounts, with loss not exceeding a small fraction of one percent in the transmission, even in the greatest distance, twelve thousand miles – to the opposite end of the globe." (Qtd in Waser)

It was suggested at the time for the powering of

electric automobiles, homes, ships, and aircraft through plasma beams and the ground over the horizon.

Tom Bearden rediscovered some basic physical properties of similar uses of dipoles noted in Tesla and Maxwell's (pre-Heaviside) work by observing various overunity experiments conducted in the last few decades by modern experimenters and inventors. Much to his surprise, physics could already explain what was occurring, with full respect to the conservation of energy, but modern electrodynamics theory is still not taught to university students as the art exists today. Using broken symmetry and Dirac Sea theory, Bearden contends, it is entirely possible to shock wires with a dipole and pull energy out of the resultant warped space-time field in the vacuum. This is the original observation Tesla made working for Edison, where he described workers being killed by the corona discharge escaping switches from the high voltage DC lines into the ground. To Tesla, this energy clearly exceeded the source and differed in nature.

Broken Symmetry was proven by Lee and Yang after its inception years earlier. According to Bearden, this explains why source charges continue to emit photons and measurable energy without decaying. In engineering applications, by maintaining voltage transfer without using current through the process mentioned above, it is theoretically possible to extend the life of dipoles indefinitely while still obtaining work. Each time the circuit is switched, however, it must be opened before the wire relaxes and current can flow. The components of the circuit are used to generate or capture this energy. Mathematically,

$P = I * V$

where voltage potential is being used and

$t = R * C$

when charging a capacitor. When

$V = I * R$ where

$V / R = 0$,

resistance and voltage are irrelevant in relation to the circuit. Depending upon capacitance, and if no current is flowing, a limited amount of charge can be obtained from nature via a radiant event, theoretically without affecting the source voltage,

$P = 0 * V = 0$.

Although this is in direct contradiction to the charge equation

$Q = t * I$,

this situation is exceptional based on experimental evidence. Current can be obtained externally to the circuit, in this case, once a circuit is opened after being saturated with radiant energy. At that point, the inductor coil oscillates, and energy is pulled into the circuit from the surrounding environment. Furthermore, a field itself can do work without being expended, for example, in a permanent ceramic magnet where

W = F x d,

F being magnetic force. A collapsing magnetic field, as illustrated in this experiment, may have even further consequences not explained by the above theory that results in the introduction of usable energy potentially exceeding the source voltage.

As you will note after performing this experiment, voltage can be created without expending any additional power beyond what you expend during switching. It can then be used for work because the derivative voltage is potential energy. This means that in a perfect world, it is theoretically possible to use the existence of a dipole for energy without expending much of the dipole itself. In the real world, however, there will be losses in the circuit, some current will flow, and not all wires are room-temperature superconductors. This will at best result in the extremely efficient use of a limited dipole, with the potential application of creating other dipoles of similar or greater magnitude. This is the function of the self-organizing nature of the universe at its most basic principle. Very few of our electrical equations in their present form even apply to this case before current begins to flow.

2.2 Theory of Operation

In order for the maximum amount of radiant energy to enter the circuit, the circuit must extend the relaxation period of the wire -the duration of time between the application of high voltage and when current begins to flow in the wires. This is accomplished by the use of an inductor and alligator clip hook-up wires in the circuit. The energy received by the circuit is believed to be directly proportionate to the mass of the conductors used.

The use of an LED not only indicates the presence of current in the wires, but also the direction and magnitude of the voltage being applied to the load (primarily through its capacitance). A small resistance (1KΩ) is used to protect the LED.

The entirety of the radiant energy absorption occurs before current begins to flow in the wires from a high voltage source. This occurs in our inductor circuit when the source voltage is removed and energy is dissipated from the breaking of the lines of magnetic force from the environmental coupling, diverted into the circuit, much like generating energy from a snapped spring.

Immediately, the voltage will exceed the equilibrium voltage in all parts of the circuit when the magnetic field begins to collapse, and a high, negative voltage spike will appear until current begins to flow in the opposite direction. When the pressure is still significant on the circuit, the energy of interest will saturate and likely oscillate any conductor or component attached to the circuit. The voltage level then returns to the equilibrium level and begins producing heat and light as current flows and energy is wasted. The precursor energy is believed to come from the environment, when the magnetic field is still attached to the Earth's magnetic field, thus the conservation of energy model is respected.

As the voltage drops below the ground, the conductors actively move current in a negative-resistance away from the ground and into the branches of the circuit, including capacitances. Thereafter, the circuit behaves normally by dissipating the energy as conventional current, even if the LED is shorted or open.

3.1 Experimental Procedure and Results

The first circuit configuration shown in *Figure 1* was assembled and connected to a 12 V DC power supply. Test leads were attached across the coil and the LED. Closing the switch built a magnetic field around the inductor which acted as our magnetic field coupling. After the voltage across the inductor balanced to zero, the switch was opened resulting in a voltage oscillation at a fixed frequency directly from the Earth's magnetic field through the coil that exceeded the source voltage, seen in *Figure 2*. When current began to flow again, the stored magnetic field began to contract, and flow through the coil, resulting in our DC current from the ground to balance out.

The experiment was then repeated with circuit configuration two with the disconnected lead at an angle and other non-parallel direction to avoid capacitance. The results were identical, even though there was not a complete circuit. This is possibly explained by capacitance in the diode. When the wire leads were removed and directly attached to the ground clips, the LED did not conduct.

The experiment was again repeated with circuit configuration three, demonstrating that the energy effect is not entirely biased to one polarity, suggesting the validity of Dirac Sea potenteation of the components.

The fourth and final configuration was chosen to demonstrate that the diode was not receiving energy through the leads as a radio or magnetic impulse through the air. The LED did not conduct.

A sunburn-like effect on the researcher was also noted. This may or may not be related to the experiment.

4.1 Implications of Results

This experiment demonstrates a simplified model of Tesla's observation of DC power lines, transmission along a single wire, and negative resistance irrefutably.

This experiment secondarily suggests that an established magnetic field, devoid of the supporting current in a coil, is capable of rotating and cutting the wires of the coil. This has generation potential, and I have already established a number of theoretical generator designs utilizing this concept.

The oscillation observed is likely the result of the magnetic field in rotation, either self-correcting its rotation as it establishes current in the wire in relation to the Earth's fixed magnetic field, or driving in suspension by another universal force in the environment, possibly an oscillating magnetic field as suggested by Stubblefield.

The observed constant rotational (or bounce) speed was recorded at approximately 11,111 rotations per second. This resulted in the 22.22 KHz oscillation.

5.1 Figures and Tables

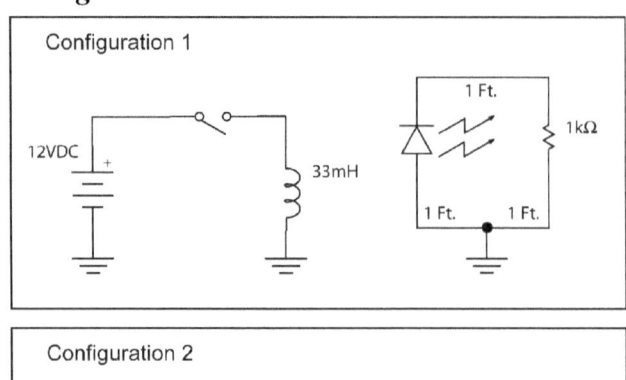

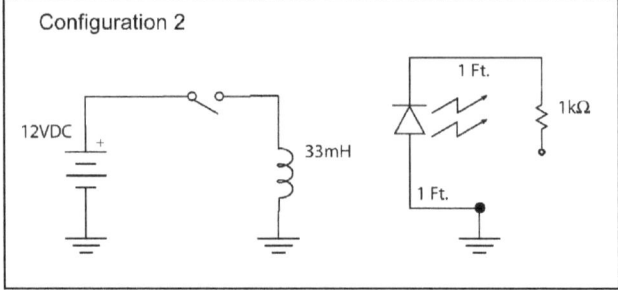

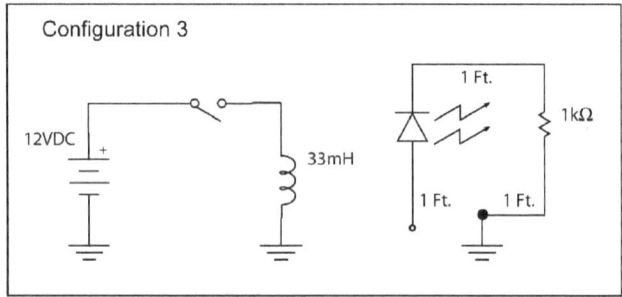

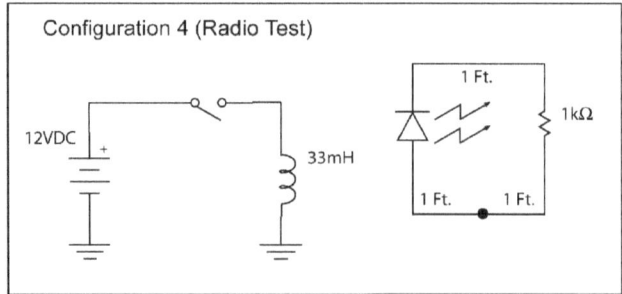

Figure 1: Test Circuit Configurations

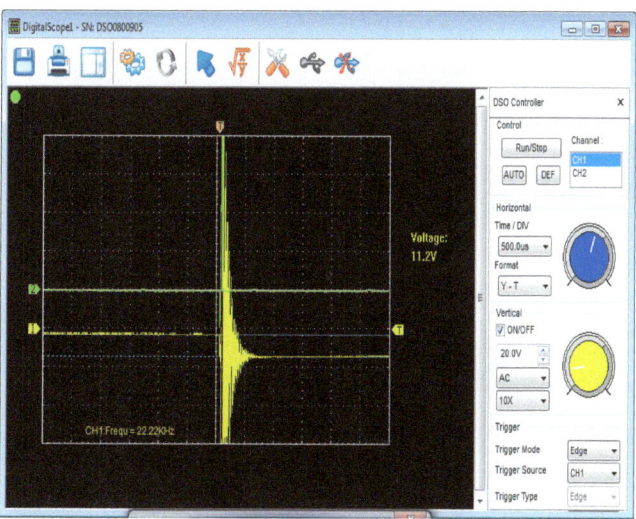

Figure 2: The radiant event causes self-oscillation in the inductor (CH1) at approximately 20KHz. Voltage is at the greatest level when the spark occurs, and is rectified by the LED (CH2)

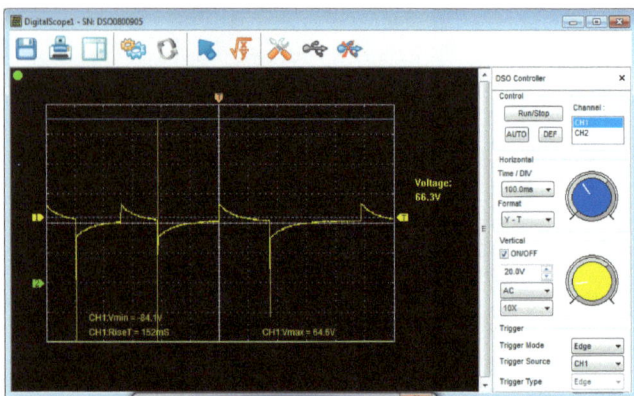

Figure 3: The negative radiant spike on the tail of the impulse exceeds source voltage by a factor of 7.

6.1 References

Waser, Andre, "Nikola Tesla's Radiations and the Cosmic Rays," *ExtraOrdianary Technology*, Vol. 1, Number 3, 2003. Tesla Tech, Inc.; 42; quoted: Nikola Tesla, "The Transmission of Electrical Energy Without Wires As a Means of Furthering Peace," *Electrical World and Engineer,* Jan. 7, 1905; 21-24.

Part 2

The Inventions

Thoughts on Magnetic Coupling

Michael C Ellis AAS, BS

November, 2019

Magnetic coupling, or as Nikola Tesla called it wireless power - in other words transmission of energy by magnetic plasmas- has remained mysterious since its inception over 100 years ago. It is a closely guarded secret that promised to overthrow the energy tyrants and save the planet. It sounds like a pretty good solution, even today!

With the PG&E wildfires, one can see how the presence of power lines are not only cumbersome and hard to maintain, but can have real consequences to people and the environment. That translates into home environments too, as one can see with the danger of leukemia, electrocution, and fire from high tension wires and smart meters, curious minds exploring outlets, unkempt appliance cordage, and faulty adapters, respectively. Disaster happens every day, no matter how safe they are made.

Is it time for free energy service? Should government invest in Earth Energy utility that runs on the Earth's magnetic field and supplies the population with un-metered energy? Like some think, the ancient Egyptians accomplished, should we take our power from the aethers and give of it freely? What would this mean for our climate crisis and Big Oil's wars across the globe to secure pipelines and eliminate dissent? If all you had to do to get power was to stick a rod in the ground and hook up a coil, what would that mean for civilization? World Peace? Let's get into the Earth coupled receiver system and see what we can do from there.

Now the technology really works in one general concept. That is of coupling the magnetosphere – usually by a coil of some type or a copper pipe in the case of orgone energy-, resonating some type of electrical component – this is usually a resistor, inductor, or diode-, and then sinking through a capacitance into a grounding element – be it a battery, orgonite, or a metallic mass of some type, usually ironous. That's all there is to it!

Tesla wanted everyone to have their own energy generators near the end of his life, as he knew the powers that be would never let a worldwide free energy system happen. Essentially, his home generator patent consisted of a raised mass of metal, a capacitor, a ground plate, a spark gap, and a coil. He called it the Radiant Energy Receiver. Now I know when I say Radiant Energy, you are likely going to go with the nay-sayers and suggest that it was only a solar-powered panel. Tesla specifically stated that his device ran at night, even stronger than during the day. The point of the panel was to couple the natural wireless power signal coming from the universe. Specifically that of the Earth.

I have made a few modifications to the original circuit. Since we don't want to cause radio chatter, we can eliminate the spark gap in exchange for a rectifying diode, which didn't exist back then. The only purpose of the spark gap, anyways, was to provide polarity. This will allow us to make our system much smaller, even desktop size, perhaps, to run a cell phone or a clock. We can ground it into a battery or a water pipe and place our plate in an attic so we don't even have to go outside or hang an annoying plate up in the air where the neighbors will see it.

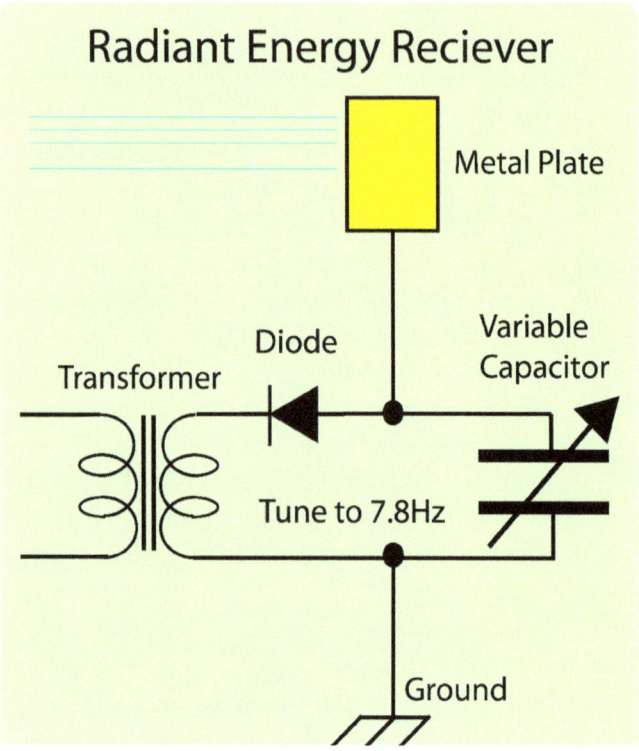

Figure 1: Radiant Energy device designed by Michael Ellis

Let's take a look at the circuit in Figure 1. As you can see, it resembles Nikola Tesla's patent very closely. The only difference is the diode, which acts to separate charges, and the variable capacitor, which allows one to tune into the Schumann Resonance. A transformer picks up the oscillations and uses them to run various appliances (not shown). It could be grounded into the positive pole of a battery if need be. Orgone generators also work. It is recommended that the plate be covered with spray paint in order to enhance the coupling effect. Remember, this energy is not electrical. It is essentially an EMP-like wave, mostly magnetic or plasmatic.

Continuing on, the next circuit I want to discuss is the Wardenclyffe-like generator or Wireless Resonator. As seen in Figure 2, this circuit is not entirely the same as the Magnifying Transmitter patented by Tesla. I have taken certain liberties to modify it for low power applications.
Now what is happening in this circuit is that the tank circuit (the combined inductance and capacitance) are oscillating when stimulated by the larger coil's signal. The larger coil is used to pick up the magnetic energy from the atmosphere. This is accomplished by using a diode to feed half of the wave into the tank. This DC impulse does not have a negative half. The severed signal creates harmonics that cause the tank to self-tune into its resonant frequency. Once the tank is resonating, it is broadcasting wireless power at the same frequency. According to the 2011 *ARRL Handbook for Radio Communications*

$$l/d = \frac{\lambda/2}{d'} = \frac{300}{2f \times d'}$$

where l is length of wire, d is diameter of wire, lambda is wavelength, d' is distance in meters, and f is frequency (resonant) in MHz. Consider this when designing your pickup coil.

To gather energy, all one has to do is design a tank receiver and pick up that wireless energy and put it to use. This circuit is basically the magnetic to electric translator.

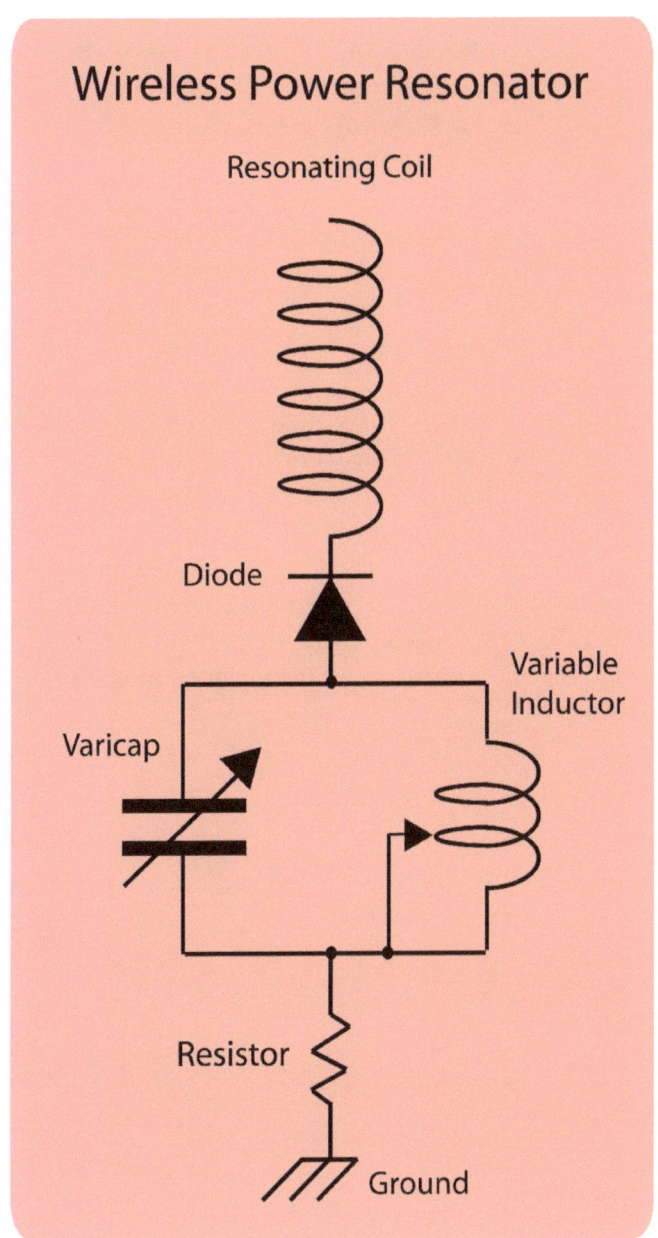

Figure 2: Wireless Power Resonator circuit designed by Michael Ellis

Another circuit I would like to add to our wireless power collection, also of my invention (with the help of some channeling), is a radiant energy receiver that works mainly off of a large coil and a capacitor in series, terminating on a battery. This is similar to the mast I derived a few years ago and put in my book *Radiant Energy: A Guide to Wireless Power.* However, this circuit is more like a Stubblefield circuit than a Tesla creation. Let's take a look at Figure 3.

Ideally, the massive coil couples the magnetic plasma, and the variable capacitor allows for tuning. The diode restricts the flow of current, limiting the energy flow to that of impulse only. Notice that the coil is shorted; that is done on purpose. Realize that true radiant energy is not a flow of current, it is actually the animation of the entire circuit all-at-once in the absence of current flow.

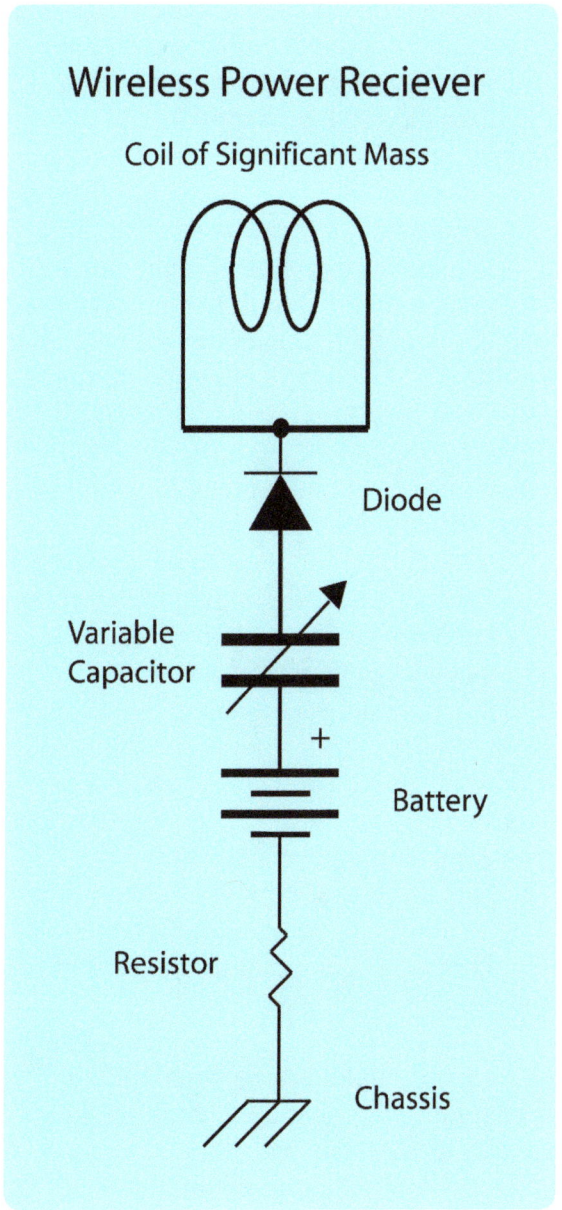

Figure 3: Wireless Power (Radiant Energy) Receiver circuit by Michael Ellis

Ideally, this should also build charge on the battery but I have not tested it yet on the oscilloscope. It does shock you, though and generate heat, and generate heat it does! I may actually classify this as a heating unit, for the practical heat it delivers.

Finally, I would like to finish with an enigma. This circuit is actually a collection of technologies that I have obtained through various means, including channeling. Essentially it uses the ground for a power source and generates power to transformer outputs.

Looking at the circuit in Figure 4, the ground is central to this circuit's function. Two tank circuits feed into an unusual transistor bridge circuit that combines the two signals. The output is rectified through a diode. The final impulses feed into a chain of transformers that animate loads. This is practically a free energy generator.

I cannot stress enough how important it is to have a solid ground connection. This device does not receive its power from the atmosphere like our other circuits do. It is of utmost importance to have a good grounding for receiving the magnetic plasma, whether that is a chassis connection or an Earth ground.

As you see, this circuit can be tuned to two different frequencies at once. That is ideal for adaptability as day switches to night, seasons change, etc. It can receive both without changes or re-tuning.

Ideally, an infinite number of loads can be added to

this. Tesla claimed this was his car circuit (his channeled spirit).

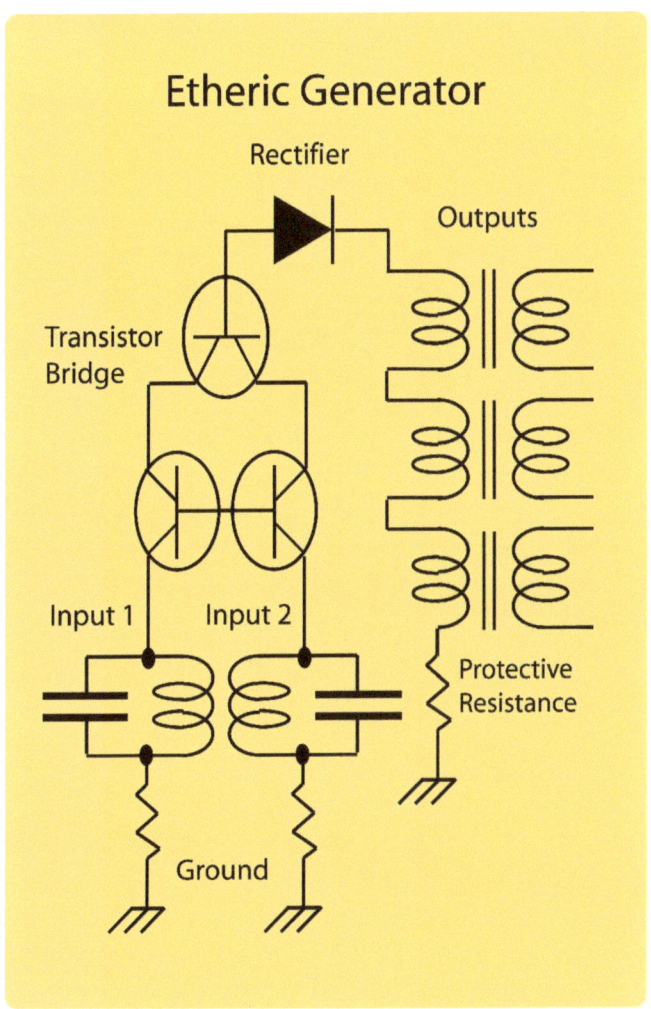

Figure 4: Ground Wave Etheric Generator by Michael Ellis

This is the kind of technology that can save our planet. We are looking at the worst technology crisis that has ever existed in the history of humanity. Humanity is at a point that we need to adapt or die - it is really that simple. Our food supply is at a crisis point, our forests are burning, our ocean life is dying, wildlife, including agricultural pollinators, are being slaughtered at an unprecedented rate. To sit by idly as the extinction of our planet occurs is to take part in its defeat. Government has been useless, greedy, and corrupt as it pertains to anything relating to protecting the planet and dealing with the climate crisis. Basically, NASA, which has known about global warming since the 80s, and the Trump administration, which is hopefully on the fast track out of office, have done little more than told the collective populace, "you're on your own." That leaves it up to the population to make the difference and take control of the situation. This is truly a war. I choose to make a stand. This is my weapon.

I would like to offer you a few more circuits I have since discovered, that will act usefully. These will work right out of the box, are scalable, and don't need much engineering to get to work. This is appropriate technology, meaning that it doesn't take much technology to make, and will fulfill a need, all while combating the use of fossil fuels. I give you The Etheric Heater and The Etheric Cooler.

These are not your traditional conditioned-air warming systems. These actually produce ambient etheric energy conditioned such that it produces a heating or cooling in the matter around it. Your body is the matter we are most concerned with. You will find that this energy is healing and at best enlightening. If it is discomforting in any way, you can simply scale it down and it will become more comforting – all things in moderation, as they say. It tends to affect humidity as well. One will notice a tropical feeling with the heater, and an arctic feeling with the cooler. I consider them both life-forces, one hot, the other cool. There is no practical way to block these energies or to steer or contain them, like orgone, they permeate everything. Cool energy will negate hot energy. Unfortunately I don't have access to a infrared camera to take photos of it, so experimentation on your end will have to do. The best option is to build a unit with both hot and cool and to alternate circuits, disconnecting the one not in use. See Figure 5 below for schematic with parts numbers.

Let me leave you with this little tidbit of advice. As Tesla once said, and I paraphrase with emphasis, once the scientists begin to study the paranormal, as I have, they will begin to learn infinitely more about the universe that they could have only sticking to their close-minded pursuits.

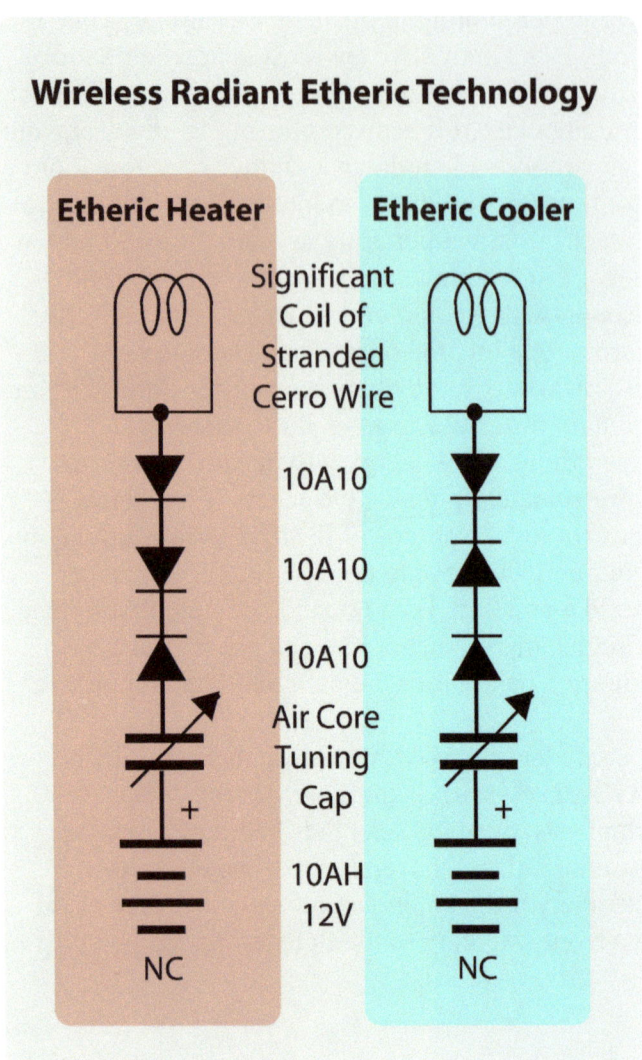

Figure 5: Heating and cooling circuits by Michael Ellis

www.ingramcontent.com/pod-product-compliance
Lightning Source LLC
Chambersburg PA
CBHW041943240526

45473CB00033B/496